ABOUT THE AUTHOR

Kathleen A. O'Grady is the originator of Authentic Intelligence™ (AQ™) and the developer of a proprietary body of work that includes the Trinity of Human Intelligence™ (THI™), The Authentic Field™ (TAF™), The Focused Field™ (TFF™), The Perception Paradox™ (TPP™), and the O'Grady Human Blueprint to Life™ (OHBL™).

Her work draws on two decades of qualitative field research across executive leadership, organizational behavior, and human development. O'Grady is the founder of Authentic Leadership Advisors, a global leadership development firm specializing in executive development, organizational culture, and large-scale behavioral change.

Authentic Intelligence: The Third Domain of Human Awareness is her first book.

She is a dual U.S. and Irish citizen.

thirddomainpress.com

KATHLEEN A. O'GRADY

Authentic Intelligence

The Third Domain
of Human Awareness

Authentic Intelligence: The Third Domain of Human Awareness

The following are trademarks of Kathleen A. O'Grady:

Authentic Intelligence™ (AQ™)
The Trinity of Human Intelligence™ (THI™)
The Authentic Field™ (TAF™)
The Focused Field™ (TFF™)
The Perception Paradox™ (TPP™)
The O'Grady Human Blueprint to Life™ (OHBL™)

All trademarked constructs remain the intellectual property of Kathleen A. O'Grady regardless of use in research, publication, or professional practice. Attribution is required in all cases. See Appendix for citation guidelines and research partnership terms.

Published by Third Domain Press
ISBN: 979-8-9949397-0-3
ISBN (ebook): 979-8-9949397-1-0
First edition, 2026.

thirddomainpress.com

Contents

Preface

The Missing Variable

Something is off.

You can feel it. A persistent sense that the faster we advance, the further we drift from one another, and the harder it becomes to agree on even the most fundamental terms of our shared survival.

Our collective intelligence, by every measurable standard, is at an all-time high.

And yet.

The tools we have built to solve problems are creating new ones faster than we can process them. The systems we have designed to connect us are deepening division.

For over a century, human intelligence has been understood through two primary domains. The first, cognitive intelligence, commonly measured as IQ, captures our ability to rationalize: to analyze, compute, plan, and solve. Second, emotional intelligence, commonly measured as EQ, captures our ability to relate: to empathize, self-regulate, connect, and communicate. Together, these two domains have shaped how we educate, how we lead, how we hire, how we govern, and how we understand what it means to be human.

This book proposes that those two domains are not enough.

There is, at minimum, a third form of intelligence operating beneath and beyond the reach of cognition and emotion. It is disciplined in its precision, rigorous in its implications, and grounded in observable human experience. It is a distinct mode of human awareness that is non-rational and non-emotional in origin, yet consistently observable in how people make meaning, sense truth, and navigate complexity when reason and empathy reach their limits.

Over the course of twenty years of applied work in human behavior, Kathleen A. O'Grady has observed a persistent perception that existing recognized forms of intelligence cannot adequately explain.

In 2022, O'Grady identified this domain and officially established it as Authentic Intelligence™ (AQ™). This book presents AQ™ for the first time in full and proposes it as a third domain of human intelligence, equal in significance to IQ and EQ, and overdue for scientific investigation.

AQ™ describes our access to resonance: the capacity to perceive, process, and respond to information through a mode of awareness that operates prior to both cognitive analysis and emotional connection.

If IQ asks "What do I think?" and EQ asks "What do I feel?", AQ™ asks a question that most people recognize but few can articulate: "What do I know, and how do I know it when thinking and feeling are not the source?"

That question has been asked before, across cultures, centuries, and disciplines. It has been filtered through philosophy, contemplative traditions, and religious narratives, then reduced to anecdotal shorthand: "gut instinct," "intuition," "inner knowing."

The real investigation starts here.

This is a theoretical proposal aimed at the scientific community. It offers a construct, a set of testable propositions, and an open invitation to take the missing variable seriously.

What if we have been accessing intelligence from only two perspectives, when it requires at least three?

I
What IQ and EQ Leave Unexplained

Consider a decision you made that you could not fully explain. One that, when you look back on it, proved to be right. Perhaps even profoundly right.

When pressed to articulate why you made this decision, you found that neither IQ nor EQ could account for the full picture. The data did not point clearly in that direction. Your emotions were not strongly pulling you there. And yet something in you recognized, with quiet certainty, that this was the correct move.

Most people, when asked about experiences like this, nod in recognition. The architect who scrapped a completed design the night before submission and delivered something he had not planned, to unanimous recognition. The physician who ordered an additional test despite no clinical indication, and found something critical. The parent who turned the car around halfway to school with no explanation, drove home in silence, and spent the rest of the day watching the news. These experiences are common, cross-cultural, and remarkably consistent in how people describe them.

They are also, within the current scientific understanding of intelligence, essentially unexplained or misclassified.

The pattern extends beyond personal decisions. Throughout recorded history, many significant leaps in human output have shared a consistent feature: they arrived ahead of any explanation. Think about works of art that captured something the culture had not yet articulated but immediately recognized. Scientific breakthroughs that preceded the experimental evidence later used to confirm them. Engineering solutions that emerged whole, with no traceable antecedent in the existing knowledge base. Literary works that named something readers had long felt but never seen articulated on the page.

These outputs arrive through a channel that skillful analysis and deep emotional engagement cannot reach. They represent a pattern of awareness that operates prior to both, producing results that neither domain, alone or in combination, can adequately explain.

IQ captures a specific and important set of abilities: structured reasoning, pattern recognition, working memory, spatial processing, verbal comprehension, and processing speed. These abilities are real, measurable, and predictive of certain outcomes. IQ scores correlate with academic performance, certain types of professional achievement, and the capacity to solve well-defined problems within known parameters.

What IQ does not capture, and was never designed to capture, is how human beings navigate situations where the parameters themselves are unknown, shifting, or fundamentally ambiguous. IQ operates within systems. It is exceptionally good at analyzing information that already exists within a recognizable structure. It is less equipped to account for a form of knowing that precedes structured analysis altogether.

This is a boundary condition, and every measurement tool has one. A telescope is an outstanding instrument, but you would not point it at the ground to understand what is beneath it. IQ is a powerful measure of rational processing capacity. The limitation is in the assumption that rational processing is the whole of intelligence.

EQ expanded the conversation about intelligence in a way that connected with millions of people who had always sensed that cognitive ability alone was incomplete.

EQ captures our capacity to perceive, understand, manage, and utilize emotions, both our own and those of others. It explains why some highly intelligent people struggle in relationships and leadership, while others with more modest cognitive profiles navigate complex social landscapes with remarkable skill. EQ introduced the language of empathy, self-regulation, social awareness, and relational competence into a conversation that had been dominated by analytical metrics for decades.

And yet EQ, like IQ, has a limit.

Emotional intelligence describes how we process and respond to emotional information. It is relational by nature. It operates in the space between people, and between a person and their own emotional landscape. It asks: "What am I feeling? What are they feeling? How do I navigate this effectively?"

What EQ does not address is the experience described at the opening of this chapter. The decision you could not fully explain, regardless of how it turned out, and that you made for reasons that could not be attributed to either analysis or feeling. That experience does not reduce to emotional intelligence any more than it reduces to cognitive intelligence. It is something else entirely.

In statistical research, when the result of an observation cannot be fully explained by the variables in the model, the unexplained portion is called a residual. It is the gap between what the model predicts and what actually happens. Residuals are signals. They tell you your model is incomplete.

The study of human intelligence is no exception to this rule.

When IQ and EQ are considered, there remains a consistent, observable thread of human awareness and decision-making that neither domain explains. It shows up across cultures, professions, and educational backgrounds. It correlates with neither domain in a way that would suggest it is a subcategory of one of them.

Researchers have circled it for decades. Studies on rapid, automatic processing have shown outcomes that conscious analysis could not replicate. Work on bodily awareness has revealed that the body registers information before the conscious mind processes it. Neuroscience has identified patterns of brain activity at rest that suggest a form of processing distinct from either deliberate thought or emotional response. Each of these lines of inquiry has revealed a piece of what appears to be a larger phenomenon.

The standard response to observations like these has been to assign them to one of the existing domains. They are attributed to unconscious pattern recognition, or to emotional attunement, or to some combination of the two. When they resist easy categorization, they are filed under anecdote, coincidence, or noise.

The evidence tells a different story.

2

Defining AQ™

We must answer the same questions any proposed domain of intelligence must answer: what is it, how does it work, and what does it explain that existing models cannot? It must do so with enough rigor to earn the attention of the scientific community, while remaining honest about what is actually being observed. That answer starts with a precise definition.

Authentic Intelligence™ (AQ™) is a domain of human intelligence characterized by non-rational, non-emotional intuitive awareness. It describes access to resonance: the capacity to perceive, process, and respond to information through a mode of awareness that operates before thinking and feeling.

AQ™

non-rational · non-emotional · pre-cognitive

Domain of human intelligence establishes AQ™ as a category of the same order as IQ and EQ, with the same claim to scientific investigation that cognitive and emotional intelligence hold.

Non-rational means that AQ™ operates outside the processes of logical reasoning, analytical thinking, and deliberate computation. It is fully compatible with reason; its origin point is simply elsewhere. A person accessing AQ™ may later construct a rational explanation for what they perceived, but the perception itself arose from a different source.

Non-emotional means that AQ™ operates outside the processes of emotional perception, regulation, and response. Emotional reactions may accompany or follow AQ™ awareness, but the awareness itself is distinct from emotion. A person may feel something about what they perceived through AQ™, but the perception did not originate in feeling.

Intuitive awareness identifies the mode of processing. AQ™ is a form of knowing that arrives whole, without the step-by-step construction characteristic of rational thought and without the affective coloring characteristic of emotional intelligence. It is direct, immediate, and often accurate in ways the person accessing it can verify after the fact, even when they could not have arrived at the same conclusion through deliberate analysis or emotional attunement.

Resonance is the operative word. Resonance implies a correspondence between a signal and a receiver attuned to that signal. AQ™ describes the capacity to receive and register information that cognitive and emotional channels are unable to detect, in the same way that certain frequencies are inaudible to the ear but measurable by instruments designed to detect them.

A definition, however clear, remains abstract until it is grounded in recognizable experience.

The most consistent description of AQ™ involves a quality of certainty that arrives without explanation. People describe it as knowing something before they have any identifiable basis for knowing it. The knowing is typically quiet rather than dramatic. It has a settling quality, as if something has clicked into place. And it is remarkably resistant to being overridden by either logic or emotion, which is often how people first notice it: they find themselves unable to dismiss a perception even when their rational mind and emotional state are both pointing in a different direction.

The language people use to describe it varies. The awareness itself is remarkably stable.

Any serious investigator encountering AQ™ will ask how it differs from concepts that occupy nearby territory.

AQ™ and gut instinct. Gut instinct captures something real, and it is often the closest language people have for AQ™ awareness before they encounter the formal concept. AQ™ may involve somatic components, but it is defined by the mode of awareness rather than its physical location. The distinction matters. When AQ™ carries significant distortion, the signal can be corrupted in transmission, producing a read that feels like AQ™ awareness but that reflects the receiver's noise rather than coherent source conditions. A gut feeling and AQ™ reception are not the same thing. Whether AQ™ awareness is reliably accurate, and under what conditions it degrades, is exactly what scientific investigation of AQ™ is designed to answer.

AQ™ and fast, automatic processing. Cognitive science has long recognized a distinction between fast, automatic thinking and slow, deliberate thinking. AQ™ shares some surface features with the fast, automatic mode: both involve rapid, non-deliberate processing. The critical distinction is that fast, automatic processing is still cognitive. It is driven by mental shortcuts, biases, and pattern matching against stored experience. It is, in essence, fast IQ. AQ™ describes a mode of awareness whose origin is neither cognitive nor emotional, and whose outputs sometimes exceed what stored experience alone could produce.

AQ™ and emotional intuition. Emotional intelligence includes a component often described as "emotional intuition": the ability to read a room, sense another person's state, or pick up on relational dynamics below the surface. Emotional intuition is relational and affective. AQ™ processes information that is neither cognitive nor affective, and it operates in contexts that have no relational component at all. A person can access AQ™ in complete solitude, about a situation involving no other people.

AQ™ and spiritual intelligence. Several researchers have proposed concepts of spiritual intelligence over the years, typically involving meaning-making, transcendence, and connection to something larger than the self. AQ™ is worldview-agnostic. It makes no claims about the source of the awareness it describes. AQ™ identifies the observable pattern. The investigation of the mechanism belongs to the scientific agenda proposed in later chapters.

AQ™ and psychic phenomena. The question of whether AQ™ and psychic phenomena describe the same territory is inevitable, and avoiding it would do a disservice to the reader. Psychic phenomena, including precognition, clairvoyance, and non-local perception, describe reported human capacities that operate outside conventional sensory channels. AQ™ does not validate or invalidate any of these claims. It occupies different ground entirely. AQ™ is a domain of intelligence, not a catalog of anomalous events. Where psychic phenomena describe what allegedly happens, AQ™ describes a mode of awareness through which alignment is recognized. Whether some reported psychic phenomena involve AQ™ awareness is an empirical question. AQ™ does not depend on the answer.

AQ™ and wisdom. The psychological study of wisdom describes a form of expert knowledge about the fundamental pragmatics of life, understood as accumulated, experiential, and developmental. AQ™ differs in a critical respect: it does not require accumulation. While AQ™ awareness may deepen with attention and practice, the capacity itself is observable in people with limited life experience, across all age groups, and in contexts where no relevant expertise exists. AQ™ appears to be an innate capacity rather than a developed achievement.

AQ™ and mindfulness. Mindfulness, as defined in contemporary research, involves present-moment, non-judgmental awareness of one's internal and external experience. Both involve a quality of attentive awareness. The distinction is that mindfulness is a practice and a state, while AQ™ is a domain of intelligence. Mindfulness may be one of the conditions that enhances access to AQ™ awareness, a hypothesis worth investigating. The two are categorically different: one is something you do, the other is something you have access to.

These distinctions point to something worth stating plainly. The pattern of awareness AQ™ describes has been recognized across virtually every major cultural, philosophical, and contemplative tradition in human history. AQ™ stands independent of all of them. The moment a construct is anchored to a particular worldview, it becomes a doctrine. AQ™ proposes a shared vocabulary, grounded in science rather than tradition, designed to be investigated across disciplines, from neuroscience and physics to psychology and information theory.

That vocabulary begins with what comes next: a model for how all three domains of intelligence function as one integrated system.

3
The Trinity of Human Intelligence™

Human intelligence has been treated as a two-channel system. One rational. One emotional. Two signals. Two ways of engaging with reality.

This chapter introduces the Trinity of Human Intelligence™ (THI™): the integrated model that describes how IQ, EQ, and AQ™ function as a single, unified system of perception and response.

The THI™ is expressed through what O'Grady identifies as the Three Rs:

We rationalize (IQ). We relate (EQ). We resonate (AQ™).

These are the three verbs of human intelligence. Every act of perception, every decision, every moment of meaning-making involves some combination of all three. Understanding how they interact, and what happens when they fall out of alignment, is central to understanding what AQ™ actually does.

We Rationalize

IQ is the structuring function within the THI™. It asks: "What is this? How does it work? What should I do about it?" It operates in the realm of the tangible: form, data, observable evidence, and verifiable outcomes.

When IQ dominates the system without the balancing influence of the other two, it produces a characteristic signature: high output, narrow aperture. In individuals, this looks like rigidity masked as logic. In institutions, it looks like optimization detached from meaning. In cultures, it looks like progress without wisdom.

We Relate

EQ is the connecting function within the THI™. It asks: "What am I feeling? What are they feeling? What does this relationship need?" It operates in the realm of the intangible: emotional data, relational texture, unspoken dynamics, and felt experience.

When EQ dominates the system without the balancing influence of the other two, it produces its own characteristic signature: high connection, low discernment. In individuals, this looks like empathy without boundaries. In institutions, it looks like consensus-seeking that avoids necessary conflict. In cultures, it looks like compassion without clarity.

We Resonate

Authentic Intelligence™ is the unifying function within the THI™. It asks a question that precedes both thinking and feeling: "What is real here, and what matters, before explanation arrives?" It operates in the realm of the pre-tangible: information that has yet to become structured enough for IQ to analyze or emotionally charged enough for EQ to process.

AQ™ orients the system. It determines what IQ and EQ should be paying attention to. Without it, rational intelligence structures whatever information happens to be in front of it, regardless of whether that information is relevant or aligned with reality. Without it, relational intelligence responds to whatever emotional signal is loudest, regardless of whether that signal reflects something real or something reactive.

Chapter Three

When AQ™ is functioning well within the THI™, rational intelligence becomes more precise because it is working with relevant information. Relational intelligence becomes more grounded because it is responding to real signals rather than noise.

When AQ™ is weak, suppressed, or overridden, the system keeps running. IQ keeps analyzing. EQ keeps relating. But the orientation is off. The system moves efficiently in directions that may have nothing to do with what actually matters. Most people recognize this condition: being busy, productive, emotionally engaged, and sensing somewhere beneath all of it that something fundamental is misaligned.

The THI™ is a system, not a taxonomy. The three intelligences are in constant interaction, each influencing the others in real time. The quality of that interaction determines the quality of perception, decision-making, and behavior.

When all three are active and communicating, the human system operates with coherence. Perception becomes clearer. Meaning becomes more accurate. Choices align with reality rather than reacting to noise.

When one or two intelligences dominate while the others are suppressed, the system loses fidelity. A person operating on IQ alone can achieve remarkable things. A person operating on EQ alone can build deep relationships. But without AQ™ orienting the system, both are operating without the calibration that tells them whether their output is aligned with what is real, relevant, and sustainable.

There is a natural sequence within the THI™, and understanding it clarifies much of what this book proposes.

AQ™ transmits first. It registers coherence or misalignment before conscious processing begins. This is the pre-cognitive signal, the felt sense of rightness or wrongness that arrives before explanation.

EQ processes second. It translates the signal into emotional and relational meaning. This is where the pre-cognitive registration becomes felt: comfort or discomfort, trust or suspicion, resonance or dissonance.

IQ processes third. It structures the signal into language, logic, and actionable understanding. This is where meaning becomes explicable: reasons, plans, justifications, narratives.

When this sequence operates cleanly, the human being perceives clearly, feels accurately, and thinks precisely. When AQ™ is active in the system, IQ and EQ have something real to work with. Without it, they risk processing in a closed loop; rational and relational signals feed each other without any coherent source orienting either.

When the sequence is reversed, which modern culture overwhelmingly encourages, IQ leads. It constructs explanations before the system has oriented. EQ reacts to the explanations rather than to reality. And AQ™, if it is consulted at all, is brought in last, asked to validate conclusions that were drawn without its essential input.

This reversal produces a recognizable sensation: the sense that something is misaligned, even though the logic checks out and the emotions feel handled. That is AQ™ signaling that the system has done its processing out of order. The orientation came after the conclusion rather than before it.

4

The Perception Paradox™

Most people assume perception starts with what they can see, measure, or name, but something has already registered before any of these senses engage.

This is The Perception Paradox™ (TPP™). It maps how meaning moves through three distinct layers of awareness on its way from potential to form. Chapter Three introduced the three layers each intelligence operates within the THI™. TPP™ reveals how awareness moves through them.

The prevailing view is that perception begins with tangible input. A stimulus must first be seen, heard, measured, or felt before the process can begin. Input, processing, understanding, action. Linear. Orderly. Rational.

The reality is different. We consistently act on information before we consciously possess it. The early perception often proves more accurate than the later analysis. Perception does not begin where we think it does.

The pre-tangible layer is where AQ™ operates within TPP™. This is where the process begins.

In the pre-tangible layer, something registers before it has taken any recognizable form. Nothing has been seen, heard, analyzed, or felt in the emotional sense. Yet the system has oriented. The person senses that something matters, that something is aligned or misaligned, that something requires attention. The sense is clear. The content is still forming.

This is the least recognized and most frequently overridden layer of perception. It is also the most foundational. Everything that follows is shaped by what happens here. If

the pre-tangible registration is received clearly, the subsequent
layers process with greater accuracy. If it is missed, suppressed,
or overridden, the subsequent layers process without
orientation, and the resulting perception, however detailed,
may be fundamentally misaligned.

The intangible layer is where EQ operates within TPP™. This is
where the process becomes felt.

In the intangible layer, the pre-tangible registration
translates into emotional and relational experience. The vague
sense that something matters becomes a feeling: comfort or
discomfort, attraction or aversion, trust or suspicion. The body
responds. The person begins to feel their way into what the pre-
tangible layer has already registered.

This is where most people locate the beginning of their
awareness. When asked how they knew something, they often
say, "I felt it." The feeling is real. In TPP™, though, the feeling is
second in the sequence, a translation of something that arrived
earlier in a form that was pre-emotional.

The intangible layer adds richness and relational meaning to
perception. It is where the abstract sense of alignment becomes
lived, as trust, comfort, or ease in the body. The insight TPP™
adds is that the emotional response translates something that
was already present before the emotion engaged.

The tangible layer is where IQ operates within TPP™. This is
where the process becomes explicable.

In the tangible layer, what was pre-tangible and then felt
becomes structured, named, and actionable. IQ organizes
perception into language, logic, categories, and plans. Evidence
is identified. Reasons are constructed. Perception becomes
something that can be communicated, defended, and acted
upon with rational confidence.

This is the final stage of a process that began two layers
earlier. It gives perception its form, makes it usable in the

shared, external world. It is also where perception is most vulnerable to after-the-fact rationalization.

When IQ processes a perception, it constructs an explanation. That explanation may accurately reflect what the pre-tangible and intangible layers registered. Or it may construct a plausible narrative that has little to do with what originally registered. Once it does so, the explanation replaces the knowing, and the person believes they arrived at their perception through logic, when in fact the logic was constructed to justify, modify, or suppress a perception that originated somewhere else entirely.

The reason this feels paradoxical is that Western culture has, for centuries, treated the tangible layer as the only legitimate form of perception. Knowing must be explicable or it is suspect. Perception must be evidenced or it is unreliable.

TPP™ reframes this assumption. The tangible layer is essential for communicating and acting on perception in the shared world, but it is the completion of a process, the arrival point rather than the origin. Requiring tangible evidence before granting legitimacy to a perception is like requiring a translated book before acknowledging that the original was written.

TPP™ also reveals where the process breaks down. Interference can enter at any layer. The pre-tangible registration can be too weak or too suppressed to orient the system. The intangible translation can be hijacked by emotional reactivity that has nothing to do with the current moment. The tangible explanation can overwrite both earlier layers with a narrative that serves comfort over accuracy.

Each type of breakdown produces a different kind of misperception, and each requires a different kind of correction. Strengthening IQ alone cannot fix a pre-tangible registration that was suppressed. Processing emotions alone cannot restore an orientation that was never received. The correction must match the layer where the interference entered.

5

The Authentic Field™
and The Focused Field™

Where does the coherence signal that AQ™ detects come from? And how does something universal become specific to one person, in one moment, facing one situation? O'Grady identifies two field constructs that answer both questions: The Authentic Field™ (TAF™) and The Focused Field™ (TFF™). Together, they describe the source of the signal that AQ™ detects, the mechanism through which that signal becomes personal, and the conditions under which signal fidelity is maintained or lost.

TAF™ is the universal field of coherence that exists prior to perception, interpretation, or localization.

Universal means that TAF™ is constant and unbounded. It is the foundational condition from which all differentiated expressions of reality emerge. It belongs to no individual, culture, discipline, or belief system. It is the context in which all of these exist.

Coherence is the condition in which the parts of something fit and work together naturally. Humans recognize it as alignment, as pieces fitting, as a situation or decision or environment feeling right in a way that precedes rational explanation. In the language of The Perception Paradox™, it is the content of the pre-tangible signal. TAF™ is its constant source.

Field is not a casual metaphor. Physics has long established that the fundamental forces governing reality operate as fields. A field is not a discrete object acting on another object. It is a continuous, dynamic condition of space itself. Electromagnetism, gravity, and quantum interactions all function this way.

O'Grady proposes that coherence also operates as a field. It is the foundational condition from which order, meaning, and potential emerge.

Prior to perception, interpretation, or localization means that TAF™ exists before differentiation occurs. It is the constant source domain containing all potential meaning and information before any of it becomes specific or embodied.

Consider the electromagnetic spectrum, which here stands in for TAF™. It exists continuously, everywhere, composed of an enormous range of frequencies. A radio receiver does not create the signal; it tunes to a specific frequency within a field that is already present. The receiver's function is to localize a portion of the field into something perceptible and usable.

TAF™ is the source field. TFF™ is the individualized localization of TAF™ where potential becomes orientable, retainable, and selectable. If TAF™ is the source, TFF™ is where the source becomes personal, the threshold where the undifferentiated potential of TAF™ becomes individualized. In the language of the electromagnetism analogy, TFF™ is the receiver. It is where "everything that could be" becomes "this."

Within the scope of this book, TFF™ is where individualized awareness lives. It is shaped by what a person pays attention to, what they have experienced, what they reinforce through repetition, and what they allow in or filter out. It is the region where universal potential meets specific expression.

This is where AQ™ becomes essential. AQ™ is what makes TAF™ accessible to TFF™. It is the intelligence that carries it from the source field into individualized awareness. Without AQ™, TAF™ and TFF™ exist in the same way that a broadcast

Chapter Five

signal and a radio exist: present, functional, yet disconnected. TAF™ is the broadcast. TFF™ is the receiver. AQ™ is what tunes it.

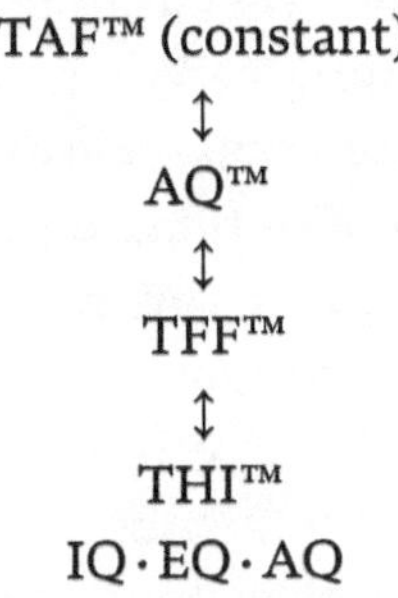

A loose analogy: a Wi-Fi signal broadcasts continuously across a space. It does not travel in a line from a router to a device. Every device within range has the capacity to receive it. But the reception varies. Walls, interference, device condition, and competing signals all affect what gets through. The broadcast never stops; what varies is reception. TAF™, AQ™, and TFF™ operate in a comparable way, with one critical difference: TAF™ has no range limitation. It broadcasts across all of existence. The only variables that affect fidelity exist at the level of TFF™.

The coherent frequency of TAF™ is infinitely accessible to all TFF™ reception via AQ™. The fidelity of that transmission is based on countless variables, most of which remain uninvestigated.

Signal fidelity bridges the field constructs with lived experience.

High fidelity means that coherence transmits from TAF™ through AQ™ into TFF™ with minimal distortion. Perception is clear. Orientation is stable. Responses are proportional to what is actually happening.

Low fidelity means the transmission is corrupted, weakened, or obscured. The person's Focused Field™ has accumulated interference that degrades the transmission. Perception is filtered. Orientation drifts. Responses may be confident, even impressive, while being fundamentally misaligned with the reality they are responding to.

Highly intelligent, emotionally competent people sometimes make decisions that seem inexplicably wrong. Their IQ is intact. Their EQ is functioning. Their AQ™ reception is compromised. The orienting intelligence that would have flagged the misalignment was either too weak to register or was overridden by the louder pull of rational certainty and emotional momentum.

If coherence is the signal, distortion is what degrades it.

Distortion is anything that disrupts the transmission loop between TAF™ and TFF™, reducing fidelity between source conditions and lived orientation. It enters the system through several channels. Fear-based filtering is among the most common: when perception is routed through survival-level reactivity, the signal from TAF™ becomes undetectable beneath the noise of defensive processing. Accumulated reinforcement of incoherent patterns is another: when TFF™ circulates the same distorted input repeatedly, that becomes its baseline. The person adapts to it and may stop recognizing it as distortion at all.

A third source is cognitive override: the condition in which IQ constructs an explanation so quickly and so confidently that it suppresses the pre-cognitive signal from AQ™ before it can register. This is the pattern TPP™ describes, where the tangible layer overwrites the pre-tangible signal. It is particularly common in cultures and institutions that prioritize rational analysis above all other forms of intelligence. The system learns to explain first and orient second, reversing the natural sequence of the THI™.

Distortion is measurable in its effects. Increased reactivity, shortened time horizons, decisions that optimize for immediate relief rather than sustainable alignment, and the progressive narrowing of what a person or system can perceive are all indicators of degraded signal fidelity between TAF™ and TFF™.

If distortion is measurable, the field generating it should be measurable too. O'Grady proposes that TAF™ functions as an empirically detectable phenomenon. Whether this is demonstrable in the physical sense is a testable claim, one that sits at the boundary of current scientific investigation. The mechanisms by which such a field might be detected and the experimental designs by which its properties might be studied are questions for the research agenda proposed in Chapter Eight.

The recognition of coherence is consistent and observable across populations. It precedes rational and emotional processing. It behaves like a signal being received rather than a conclusion being constructed.

The question that follows is practical: what does this pattern look like when the construct is applied to recognizable human experience?

6

AQ™ in Operation

Every signal needs a context to register. The following examples are drawn from patterns documented across two decades of direct observation, mapped here against the construct for the first time.

The Decision That Precedes Its Own Explanation

Chapter One opened with a familiar experience: the decision that proved right before the person who made it could say why. Now that the full construct has been defined, the mechanics become legible.

A senior executive is presented with an acquisition that meets every rational criterion: sound financials, a clear strategic fit, and an advisory team unanimously in favor. EQ reads the room: the energy is high, the momentum is forward, the relationships are aligned. IQ and EQ are both signaling yes.

The executive pauses. Something has not settled. There is no objection she can name, no data point she can cite, no relational cue she can identify. The pause itself is the signal. Authentic Intelligence™ has registered something that neither IQ nor EQ has detected.

Coherence from TAF™ is transmitting through AQ™ into her Focused Field™, and what is arriving does not confirm what the rest of the system is reporting. The pre-tangible layer of TPP™ has picked up something the tangible and intangible layers have not yet reached.

If she overrides the pause, she will reverse the natural sequence of the THI™. IQ will lead. EQ will follow. AQ™ will be suppressed. The decision may still succeed on its own terms. But if the dissonance was real, the cost will surface later, in ways that rational analysis and emotional intelligence will struggle to explain after the fact.

If she honors the pause, the THI™ will operate in sequence. The decision may be delayed. It may change entirely. It will be aligned.

The Creative Arrival

The first chapter noted a recurring signature across many creative and scientific breakthroughs. The third domain clarifies why this happens: how something arrives whole, before analysis has engaged, and why that arrival is not an accident or anomaly but the natural function of AQ™ operating without interference.

A composer sits at the piano and a phrase emerges that she did not plan, did not build from theory, and cannot trace to any prior influence. It arrives whole. She recognizes it immediately as right, not because she has evaluated it, but because something in her registered its coherence before evaluation began.

There is a quality to the moment that creators across disciplines describe in remarkably similar terms: a sense of urgency that is not anxious, a clarity that the material itself is pressing forward, asking to exist. The composer becomes less the author and more the conduit. The phrase is not written for her or even for the listener. It is written through the space between them, toward what wants to become possible by way of that space. The creation serves something that neither creator nor recipient could have specified in advance, yet both recognize immediately on arrival.

In the language of the third domain, TAF™ supplies
coherence continuously. AQ™ carries it into TFF™. When
TFF™ is clear, distortion is low, and the signal path is neither
obstructed nor oversaturated by IQ and EQ, what arrives
can exceed what the person's accumulated knowledge would
predict. The composer did not generate the phrase from stored
knowledge. She received it through a transmission pathway that
her training, her stillness, and her attention had kept open.

The same dynamic appears in scientific insight. A
researcher working on a question for months suddenly
sees the solution in a moment that has nothing to do with
the question itself. The insight arrives through a shift
in orientation that allows what analysis alone could not
assemble to arrive.

IQ will later structure the insight into publishable form. EQ
may register the thrill of discovery. AQ™ delivered the signal.

The Relational Registration

Some people walk into a room and register something about
another person before a word is spoken. This is commonly
attributed to emotional intuition, to EQ reading non-verbal
cues and relational dynamics below the surface. In many cases,
that attribution is accurate. EQ is extraordinarily capable in
relational contexts.

But there is a category of relational knowing that EQ
does not account for. The immediate sense that a person is
trustworthy or that something about them is incongruent,
based on something prior to emotional content entirely. No
warmth, no coldness, no visible cue.

In this case, AQ™ detects coherence or distortion in
another person's Focused Field™ before EQ engages with the
relational data. It is pre-tangible and pre-emotional, operating
across differentiated fields at the level of coherence rather
than personality.

This is why AQ™-based recognition often resists revision. Emotionally derived impressions update as new relational information arrives. AQ™-based recognition tends to persist, because it is picking up something more foundational than behavior or emotional presentation.

The registration did not require proof to form, though the person may later find it.

The Override

The previous examples describe single moments of recognition. But what happens when AQ™ awareness is repeatedly ignored?

A person spends years in a career that satisfies every external measure.

The role is respected, the compensation is strong, and the relationships are functional. IQ confirms the logic: this is a sound position. EQ confirms the relational investment: people depend on you here.

Beneath both, a persistent signal has been transmitting all along. Easily mistaken for dissatisfaction or burnout, it is the steady, quiet awareness that the life being built does not correspond to the life pressing to emerge.

It persists, and each time it is overridden the cost is incremental: energy narrows, creativity contracts, and decisions that once felt easy begin to require effort. The gap between performance and orientation widens so gradually that the person adapts to it, mistaking the adaptation for maturity.

This is distortion accumulating in TFF™. The signal from TAF™ has not stopped. AQ™ has not failed. The transmission pathway has been narrowed by years of reinforcement in a direction that IQ and EQ validated but AQ™ never confirmed.

From the outside, the decision to leave appears sudden, shocking, maybe even impulsive or irresponsible. From

the inside, it is the moment the signal finally overrides
the blockage and becomes embodied. It is often felt as an
overwhelming sense of relief.

What Becomes Visible

The same loop described in the previous chapter is operating
in each of these contexts. What varies is fidelity. The executive
who pauses before the acquisition, the composer who receives
the phrase, the individual who registers incongruence in a
stranger, the person who walks away from the career that
checked every box except coherence: each is receiving the
same signal through the same transmission pathway, with
varying degrees of clarity.

7

The Domain Silo Effect

If AQ™ describes something very real yet intangible, something documented as far back as the beginning of written language itself, why has it been so consistently overlooked by researchers? Because it suffers from the same issue as most IQ-governed structures: competing silos.

The Structure

The fragmentation of information is species-wide.

In corporations, rational intelligence creates departments, reporting structures, standard operating procedures, and chains of command. These are useful. They are also isolating. Subject matter experts are separated from one another. Communication slows and degrades through layers of approval, translation, and hierarchy. Meanwhile, relational intelligence within those same structures introduces competition. Departments protect territory. Personalities collide. What begins as organizational design produces toxic cultures, internal politics, and the kind of noise that makes coherent decision-making nearly impossible.

In governments, political parties, religious institutions, communities, and families, the pattern repeats at every scale.

IQ establishes labels, roles, responsibilities, rules, laws, and the conditions for abuses of power. These structures are intended to create order. They also create rigidity, gatekeeping, and systems that resist the very information that could improve them. EQ, operating within those rigid structures, produces individuation, tribalism, and escalating cycles of us-versus-them. Fear becomes the operating system. The results are visible across history and are accelerating in the present: catastrophes of human rights, exploitation of the natural world, and an increasing inability to act collectively on threats that require collective response.

The Sensing

AQ™ is what senses the misalignment, that the structure has become more important than the coherence it was built to serve. Whether in a boardroom, a government, a family, or an individual life, the sensing is consistent and persistent. But without a way to validate it, without language, without institutional support, without even a name, it is ignored or overridden by everything listed above, and much more.

This is the domain silo effect: a structural consequence of building systems on two intelligences while suppressing the third.

The Developmental Stage

Most of what exists in nature is self-organizing. Cells repair. Ecosystems balance. Bodies regulate without instruction. This is coherence operating without interference. IQ and EQ are extraordinary intelligences that have reached a level of sophistication now capable of recognizing what preceded them. This is a developmental stage. The structures IQ built and the relationships EQ navigated were necessary steps toward the threshold where AQ™ could become visible.

The cost of reaching that threshold without AQ™'s orienting function is now measurable: optimizing without alignment, connecting without discernment, and accelerating without coherence. The results are visible across every human institution and increasingly across the ecosystems they depend on.

The Disciplinary Blind Spot

The disciplinary blind spot is worth examining in detail, because
the university is the institution most responsible for determining
what counts as knowledge, and therefore what gets studied.

The scientific method is a product of IQ. It operates through
hypothesis, controlled observation, measurement, replication,
and logical analysis. It requires variables to be isolated,
conditions to be controlled, and conclusions to be derived
from evidence that can be independently verified. These
are extraordinary strengths. They have produced the most
significant expansion of human knowledge in the history of
the species.

The organizational structure that grew up around the
scientific method reflects its logic. Knowledge was divided into
disciplines. Each discipline defined its scope, its methods, its
standards of evidence, and its boundaries. Physics studied matter
and energy. Chemistry studied molecular interaction. Biology
studied living systems. Psychology studied the mind. Philosophy
studied meaning.

This structure works brilliantly for questions that fit within a
single discipline. It works less well for those that span multiple.
And it is structurally ill-equipped to detect phenomena that
exist between them, or beneath them, or across all of them
simultaneously. AQ™ is precisely that kind of phenomenon.

The Researcher Dilemma

Consider how a researcher might encounter the phenomenon of
awareness AQ™ describes.

Neuroscience. Studies of resting-state brain activity have
surfaced evidence suggesting a form of processing distinct
from deliberate thought or emotional response. The finding is
publishable and confined to neuroscience. It stays within the
department.

Physics. A physicist studying quantum coherence notices parallels with descriptions of intuitive awareness, but pursuing that parallel falls far outside what the department would support.

Psychology. A psychologist documents cases in which people make accurate judgments that exceed what their available information could support, files the anomaly, and moves on.

Information theory. A theorist modeling signal transmission and noise encounters signs that suggest a form of pre-cognitive filtering operating prior to conscious processing, but the finding sits outside an established disciplinary home.

Each encountered a piece of what AQ™ describes, produced legitimate findings within their discipline, and was structurally prevented from seeing the larger picture, because it spans all of them and belongs exclusively to none.

IQ built a structure that can only detect what its own methods were designed to find. AQ™ does not conform to those boundaries. It crosses disciplines. It operates prior to the kind of conscious processing that controlled experiments are designed to capture. It involves a form of awareness that resists isolation into single variables. It produces outcomes that are observable but difficult to replicate on demand under laboratory conditions, because the conditions it requires cannot be manufactured in a lab.

The Self-Reinforcing Structure

The structure is self-reinforcing. Departments hire researchers trained in the department's methods. Those researchers train the next generation in the same methods. Journals publish work that meets the department's standards of evidence. Funding follows established research agendas. Tenure rewards specialization. Cross-disciplinary work is acknowledged in principle and penalized in practice, because it is harder to publish, harder to fund, and harder to evaluate by reviewers who are themselves specialists. Interdisciplinary institutes exist. The language of collaboration is ubiquitous. And yet the way we evaluate knowledge, credential researchers, allocate resources, and define what counts as evidence remains IQ-dominant.

The same pattern that fragments corporations, governments, and families also fragments the institution responsible for studying intelligence itself.

What Investigation Requires

If this explains why AQ™ has been missed, it also dictates what investigating AQ™ requires. Researchers willing and equipped to work across the boundaries of neuroscience, physics, information theory, and the philosophy of mind. New methodological approaches that can accommodate a pattern of awareness that operates prior to conscious processing. Institutional support for research that does not fit neatly into existing departmental structures.

The history of science is, in part, the history of discoveries that required new methods because the existing methods were designed for different phenomena. Quantum mechanics required new mathematics. Relativity required new geometry. The investigation of AQ™ may require a new form of inquiry altogether.

8

A Scientific Agenda for AQ™

AQ™ raises questions that reach across all disciplines. No single starting point is obvious, which is itself a sign that the construct belongs to no existing silo. What follows are entry points, not boundaries.

The Neurological Question

Is there a detectable neural signature associated with AQ™ awareness? The pre-tangible layer of TPP™ describes a form of registration that occurs before conscious cognitive or emotional processing. If this registration is real, it should leave a trace. Neuroimaging technologies, particularly high-resolution EEG, MEG, and fMRI, may be capable of detecting patterns of brain activity that correspond to pre-cognitive orientation.

Research into the default mode network, temporal lobe activity, and pineal-adjacent regions has already produced findings suggestive of processing that falls outside ordinary waking cognition. The question is whether these findings, investigated under the AQ™ construct, point to a single domain of intelligence.

The Quantum Cognition Question

Does the coherence AQ™ transmits operate according to principles consistent with quantum field theory? This is the most speculative of the questions AQ™ raises, and it must be approached with rigor precisely because the intersection of quantum physics and human awareness has attracted significant pseudoscientific noise.

The question is legitimate: quantum coherence is a well-established physical phenomenon, and the proposal that a universal field of coherence supplies the conditions AQ™ detects is, at minimum, consistent with the physics of how fields operate. Whether this consistency extends to a demonstrable mechanism is an empirical question. Computational modeling may provide an early pathway for testing whether TAF™-like field dynamics can produce the kind of signal behavior AQ™ describes.

Existing theories occupy adjacent territory. Integrated information theory proposes that awareness correlates with the degree of integrated information in a system. Neurophenomenological approaches propose that first-person experience and neural activity must be studied together. Each of these touches on ground that AQ™ occupies. None of them accounts for it fully. Whether AQ™ provides a basis for resolving what these theories leave open is itself a testable proposition.

The Somatic Question

Does the body register coherence and distortion in measurable ways prior to conscious awareness? O'Grady's work proposes that AQ™ signal fidelity has somatic correlates: that coherence registers as expansion, ease, and alignment, while distortion registers as constriction, tension, and disconnection.

Somatic measurement methods, including heart rate variability, galvanic skin response, and electromyography, may be capable of detecting these states with enough precision to distinguish them from standard emotional or stress responses. The critical research design challenge is isolating pre-cognitive somatic responses from the emotional

responses that follow them, a distinction that maps directly
onto the pre-tangible and intangible layers of TPP™.

The Cross-Cultural Question

Is AQ™ awareness consistent across populations? Formal
investigation would involve structured phenomenological
research across diverse populations, documenting the
qualitative features of pre-cognitive awareness and testing
whether they exhibit the consistency the work predicts.

This line of inquiry would also test the worldview-agnostic
claim: that AQ™ describes the same pattern of awareness
regardless of the cultural, spiritual, or philosophical lens
through which a person interprets it.

The Information Theory Question

Can AQ™ be understood as a signal-processing function
within a broader information system? AQ™ operates as the
transmission of coherence from TAF™ through TFF™ into
individualized awareness, with fidelity as the key variable.
This maps naturally onto information theory: a source, a
receiver, signal, noise, and fidelity.

If AQ™ operates within measurable constraints, what
are the bandwidth limits? What conditions increase or
decrease signal-to-noise ratio? What constitutes the noise
that degrades the signal? Information-theoretic modeling
may provide a quantitative basis for what this work describes
qualitatively.

Methodological Challenges

Investigating AQ™ will require methodological innovation.
Portions of the standard experimental toolkit will be directly
applicable. Others will need to be adapted, because AQ™

operates in ways that challenge several default assumptions of
conventional research design.

Timing. AQ™ operates prior to conscious processing. Self-
report measures, the backbone of psychological research,
capture the signal only after it has been translated through the
intangible and tangible layers. Research designs will need to
prioritize physiological and neurological markers that can detect
the signal in real time, before subjective interpretation begins.

Isolation. Experimental designs will need to create
conditions that allow researchers to distinguish pre-cognitive
orientation from unconscious cognition (fast IQ) and pre-
conscious emotional response (fast EQ). Solving this may
require new paradigms rather than adaptations of existing
ones.

Replication. AQ™ awareness is widely reported but difficult
to produce on demand under controlled conditions. Research
designs may need to shift from attempting to produce AQ™
awareness in the laboratory to identifying and studying it as
it occurs naturally, using ecological momentary assessment,
wearable physiological monitoring, and other methods that
capture real-time data in real-world conditions.

Interdisciplinary. No single methodology will capture AQ™
fully. Neuroimaging captures the brain. Somatic measurement
captures the body. Phenomenological research captures lived
experience. Computational modeling captures the dynamics.
Physics captures the field-level question. A research program
adequate to AQ™ will need to integrate all of these.

Institutional Environment

The foundational investigation of AQ™ belongs in the hard
sciences. It must happen where the tools are precise enough,
the standards are demanding enough, and the institutional
credibility is high enough to produce findings the broader
scientific community will take seriously.

                                              Chapter Eight

Several types of institutions are positioned to support this work. University-affiliated interdisciplinary research centers focused on cognition or complex systems. Independent research institutes with cross-disciplinary mandates. Programs that already bridge neuroscience and physics, such as those investigating quantum biology or integrated information.

Expected Findings

A research program built around AQ™ would, if successful, produce several categories of findings.

- Empirical evidence that a form of awareness exists distinguishable from both cognitive and emotional processing. This is the foundational claim, and it is testable.
- A clearer understanding of the conditions that enhance or degrade AQ™ signal fidelity. The factors that produce distortion, including fear, cognitive override, and accumulated incoherent reinforcement, should be identifiable and their effects measurable.
- A contribution to the study of human awareness itself, introducing a dimension that existing models have not yet accounted for.
- A potential bridge between the physical and experiential study of coherence. If TAF™ operates as a field, and if AQ™ operates as a transmission function within that field, the implications for physics, information theory, and the philosophy of mind are substantial. Of the four, this would take the longest to achieve, and it carries the most transformative potential.

9
What Changes When AQ™ Is Understood

We began with a simple observation: something is off.

We are getting smarter and more emotionally literate. And yet, by nearly every meaningful measure, we are losing our ability to orient, to perceive clearly what is real, what matters, and what to do about it.

Eight chapters later, the construct is in place. AQ™ has been defined. The perceptual structure has been mapped. The field constructs have been proposed. A scientific agenda has been laid out. What remains is the question everything has been building toward: what actually changes when AQ™ is understood?

The answer begins with science. It ends somewhere much larger.

What Changes for Science

When the research agenda detailed in the previous chapter is pursued, the implications cascade. Neuroscience gains a new investigative target. Psychology gains a model for the residual in human decision-making that cognitive and emotional models have never explained. Physics gains a testable proposition about coherence as a field phenomenon. The philosophy of mind gains a specific claim about a form of awareness that is prior to and distinct from both cognition and emotion.

Together, these represent a potential reorganization of how the study of human intelligence is structured, moving from departmental silos toward the kind of integrated, cross-disciplinary inquiry that the phenomenon demands. Pre-cognitive awareness is no longer anecdote; it is a researchable scientific inquiry. The science of perception becomes more complete, more faithful to what it claims to study, and more capable of scientific inquiry.

What Changes for Technology

AQ™ changes the terms of the conversation about artificial intelligence (AI).

The current trajectory of AI development is the acceleration of IQ; that is, the rational processing function. Machine learning systems process information faster, identify patterns more efficiently, and generate output at a scale and speed that human cognitive processing cannot match. This is impressive. It is also, from the perspective of the THI™, profoundly one-dimensional.

AI, as it currently exists, rationalizes. It can simulate relational output with increasing sophistication, yet the pre-tangible layer remains entirely beyond its reach. It operates within the tangible domain: form, structure, and measurable output.

When AQ™ is understood, the conversation shifts from "Will AI replace human intelligence?" to "What dimensions of human intelligence does AI not have access to, and why do those dimensions matter?"

AI can accelerate the tangible layer indefinitely without ever touching the pre-tangible layer. It can optimize what humans think about without influencing what humans orient toward. As AI capabilities expand, the human capacity that will become most essential, rather than most obsolete, is the one that operates in the domain AI cannot reach: AQ™.

The urgency of understanding AQ™ is therefore directly proportional to the speed of AI development. The faster AI advances, the more critical it becomes for human beings to understand and strengthen the form of intelligence that remains uniquely ours.

The pattern has precedents. When calculators arrived in classrooms, students were told not to rely on them. Within a generation, reliance was universal, and largely harmless, because mathematics is closed. The answers are verifiable. Computers extended the same trust into information management, where the structure was still visible and the logic still traceable.

AI breaks the pattern. It operates in open territory: generating meaning, simulating reasoning, producing output that reads as authoritative regardless of whether it is accurate. The tool has moved from a domain where answers can be checked to a domain where discernment is the only filter. And discernment is precisely what most users have been trained, by decades of reliable machine output, to outsource.

A generation raised on calculators learned to trust the machine for math. A generation raised on AI is learning to trust the machine for meaning. The difference is that math will correct you. Meaning will not.

This is where AQ™ becomes essential to the species, not as theory but as function. IQ can evaluate an argument after it arrives. EQ can register whether it feels right. AQ™ is what senses whether the signal is coherent before either one engages. Without it, the user has no way to distinguish between information that is fluent and information that is real.

What Changes for Shared Reality

AQ™ changes what is possible.

IQ does not reach this territory. IQ operates within frameworks, and the frameworks themselves are what is contested. Adding more rationality to a dispute between competing rationalizations produces more sophisticated arguments, not more resolution.

EQ does not reach it either. EQ operates within relational groups, and the relational groups themselves are what is fractured. Empathy flows freely within tribes and stalls at tribal boundaries. Increasing emotional intelligence within a group that already agrees with itself produces deeper in-group bonding, not broader cross-group understanding.

AQ™ offers something neither can provide: a form of awareness that operates beneath ideology, beneath identity, beneath the frameworks and relational structures that divide. Coherence, as described by TAF™, is universal. It is recognizable across every boundary that currently separates human beings from one another.

AQ™ proposes something more fundamental than the elimination of disagreement: a shared perceptual substrate beneath the disagreements. A frequency that every receiver can tune in to, regardless of what other stations they are listening to. Shared reality is recoverable, because the source of shared reality was never the frameworks, it was the coherence beneath them.

The Root

Zoom out far enough, and a pattern becomes visible. Political polarization. Ideological rigidity. Ecological degradation accelerating despite overwhelming evidence. Information systems that flood the species with noise while starving it of signal.

From the surface, these appear to be separate conditions. From the perspective of AQ™, they share a single root: fear.

Fear is the master distortion. It is the interference pattern that degrades every other signal. When fear saturates a person's Focused Field™, AQ™ transmission collapses. When fear saturates a culture, collective AQ™ collapses. When fear saturates a species, the orienting intelligence that would allow it to perceive clearly, act coherently, and respond to reality rather than to its own projections goes offline.

What is left is a species running on IQ and EQ alone. Rationalizing without direction. Relating without signal.

Building faster, reaching further, and drifting more with every cycle.

The Counterweight

AQ™ is the counterweight to fear. It is the intelligence that operates prior to the survival response, that detects alignment before the defensive system activates, that orients perception toward what is real rather than what is threatening.

When an individual strengthens AQ™ access, the fear response remains, but it loses its monopoly on perception.

When a culture begins to value AQ™, institutions begin to orient before they optimize. Governance begins to seek coherence rather than consensus. Technology begins to be guided by wisdom rather than capability alone.

When a species understands AQ™, the possibility becomes clear: human beings operating with all three domains of intelligence, integrated and in sequence. Orientation restored. The signal leading, and the noise losing its grip.

The investigation does not end here. If coherence operates as a field, its implications extend beyond human intelligence, beyond the species, beyond the boundaries this book has maintained. That investigation is next.

We rationalize. We relate. We resonate.

We always have.

Kathleen A. O'Grady
Originator, Authentic Intelligence™ (AQ™)
thirddomainpress.com

Glossary, Citation Guidelines, and Research Partnership

PART ONE: Glossary of Trademarked Constructs

The following terms are proprietary constructs developed by Kathleen A. O'Grady. Each is trademarked and should be cited accordingly in any research, publication, or professional use. The recommended citation format appears in Part Two of this appendix.

Authentic Intelligence™ (AQ™)
- **Classification:** Unifying intelligence / transmission capacity
- **Definition:** The non-rational, non-emotional intelligence that operates prior to and with IQ and EQ. AQ™ is the capacity to recognize coherence, misalignment, and significance before cognition and emotion engage.
- **Function:** Transmits coherent source conditions from TAF™ into TFF™. Enables humans to orient perception and response from resonance rather than rationalization or emotional reaction. Integrates the Trinity of Human Intelligence™ in real time.
- **Domain placement:** Third intelligence in the THI™ (We Resonate). Transmission mechanism between TAF™ and TFF™. Pre-tangible layer of TPP™.

The Trinity of Human Intelligence™ (THI™)
- **Classification:** Perceptual model
- **Definition:** The integrated system of three intelligences humans use to process and respond to reality: IQ (rational thinking), EQ (relational awareness), and AQ™ (resonance).

- **Function:** Provides the perceptual lens through which coherence becomes lived experience. When integrated, IQ structures information, EQ relates to experience, and AQ™ orients the system so both remain aligned with what is sustainable and real.
- **Expressed as the Three Rs:** We Rationalize (IQ). We Relate (EQ). We Resonate (AQ™).

The Authentic Field™ (TAF™)

- **Classification:** Source field
- **Definition:** The universal field of coherence that exists prior to perception, interpretation, or localization. TAF™ is the constant source domain containing all potential meaning and information before it becomes specific or embodied.
- **Function:** Supplies stable, coherent source conditions continuously. Maintains order as the foundational condition from which all differentiated fields emerge.

The Focused Field™ (TFF™)

- **Classification:** Localized field / threshold
- **Definition:** The individualized localization of The Authentic Field™ where potential becomes orientable, retainable, and selectable. TFF™ is awareness narrowed through attention and focus.
- **Function:** Acts as the threshold where transmitted coherence from TAF™ becomes noticeable, relational, and meaningful to an observer. Retains and reinforces orientation through repetition, amplification, and circulation.

The Perception Paradox™ (TPP™)

- **Classification:** Perceptual model
- **Definition:** The model describing how relevance forms before meaning becomes tangible, moving through three layers of perception: pre-tangible (AQ™ domain), intangible (EQ domain), and tangible (IQ domain).

- **Function:** Shows that meaning is incubated in the pre-tangible layer, becomes emotionally and behaviorally active in the intangible layer, and manifests as observable form in the tangible layer. Explains how perception operates prior to conscious recognition.

Supporting Terms

Coherence The state in which the parts of something fit and work together naturally without effort or force. The signal quality that AQ™ transmits from TAF™. What humans recognize as "things making sense together."

Distortion Anything that disrupts coherence in transmission, perception, or embodiment. The interference pattern in the signal loop between TAF™ and TFF™. Occurs when perception is filtered through fear, defense, or survival-based reactivity. Accumulates in TFF™ through repetitive reinforcement of incoherent patterns.

Resonance The felt recognition of coherence. The immediate, pre-cognitive sense that something fits, aligns, or makes sense. The experiential signature of AQ™ in operation. The "We Resonate" function in the THI™.

Signal Fidelity How accurately coherence transmits from TAF™ through AQ™ into TFF™. High fidelity means clear reception with minimal distortion. Low fidelity means the signal is corrupted, weak, or obscured by interference.

Differentiated Fields Distinct, localized expressions of The Authentic Field™, including people, systems, objects, activities, and states of being. Differentiation creates the possibility of relationship, interaction, and perception. The "many" emerging from the "one."

Rational Intelligence (IQ) The intelligence used for thinking, reasoning, analyzing, and explaining. First R in the THI™ (We Rationalize). Tangible-layer intelligence in TPP™.

Relational Intelligence (EQ) The intelligence used for emotion, empathy, and interpersonal connection. Second R in the THI™ (We Relate). Intangible-layer intelligence in TPP™.

PART TWO: Citation Guidelines

Authentic Intelligence™ (AQ™) and all associated constructs are the proprietary intellectual property of Kathleen A. O'Grady. Any use of these constructs in research, publication, educational materials, or professional practice requires proper citation.

- **Recommended citation format:** O'Grady, K. A. (2026). *Authentic Intelligence: The Third Domain of Human Awareness.* Third Domain Press.
- **In-text citation:** (O'Grady, K. A., 2026)
- When referencing specific constructs, the trademark symbol (™) should be included at first use within any document. Subsequent references within the same document may use the acronym without the symbol.
- **Example first use:** "The Trinity of Human Intelligence™ (THI™), as proposed by O'Grady, K. A. (2026), describes the integrated system of IQ, EQ, and AQ™."
- **Example subsequent use:** "The THI™ describes the perceptual lens through which coherence becomes lived experience."

Researchers, educators, and practitioners are encouraged to engage with and build upon these constructs. Attribution to Kathleen A. O'Grady as the originator is required in all cases.

PART THREE: Research Partnership

The investigation of Authentic Intelligence™ is an open and active invitation. Kathleen A. O'Grady welcomes collaboration with researchers, institutions, and interdisciplinary teams interested in the formal scientific study of AQ™ and its associated constructs.

Scope of collaboration. Research partnerships may include empirical investigation of AQ™ phenomena, development of assessment instruments, neuroimaging and psychophysiological studies, computational modeling of TAF™ and TFF™ dynamics, cross-cultural phenomenological research, and theoretical development of the AQ™ domain for scientific publication.

Authorship. Any research that investigates, tests, or extends constructs originated by Kathleen A. O'Grady should include O'Grady as co-author or, at minimum, provide formal attribution as the originator of the theoretical domain. Authorship arrangements will be established in writing prior to the commencement of any collaborative research.

Intellectual property. All trademarked constructs (AQ™, THI™, TAF™, TFF™, TPP™, OHBL™) remain the intellectual property of Kathleen A. O'Grady. Research partnerships do not transfer ownership of any constructs. Derivative works, adaptations, and extensions should be discussed and agreed upon in advance.

Licensing. Organizations seeking to incorporate AQ™ constructs into training programs, assessment tools, curricula, or commercial products should contact Kathleen A. O'Grady to discuss licensing arrangements.

Contact. All research partnership inquiries, citation questions, and licensing requests should be directed to thirddomainpress.com.